VANISHING OZONE

**OTHER SAVE-THE-EARTH BOOKS
BY LAURENCE PRINGLE**

LIVING TREASURE:
Saving Earth's Threatened Biodiversity

OIL SPILLS:
Damage, Recovery, and Prevention

A SAVE-THE-EARTH BOOK

LAURENCE PRINGLE

VANISHING OZONE

Protecting Earth from Ultraviolet Radiation

MORROW JUNIOR BOOKS
New York

Permission for the following photographs is gratefully acknowledged: American Academy of Dermatology, p. 22; Bureau of Reclamation, U.S. Department of the Interior, p. 6; E. I. Du Pont de Nemours & Company, p. 53; National Aeronautics and Space Administration (NASA), pp. 16, 31, 32, 35, 37; National Oceanic and Atmospheric Administration (NOAA), p. 10; National Science Foundation (NSF), pp. 28, 36, 46, 47, 51; National Weather Service, p. 54; Standard Oil, p. 27; Susan Solomon, p. 34; U.S. Geological Survey, p. 42; University of California, Irvine, p. 20. All other photographs by the author.

Illustrations on pp. 14 and 21 by Ann Neumann.

The text type is 13-point Trump Mediaeval.

Design by Trish Parcell Watts

Printed in the United States of America.

1 2 3 4 5 6 7 8 9 10

Library of Congress Cataloging-in-Publication Data
Pringle, Laurence P.
Vanishing ozone/by Laurence Pringle.
p. cm.—(A Save-the-Earth book)
Includes bibliographical references and index.
ISBN 0-688-04157-4 (trade)—ISBN 0-688-04158-2 (library)
1. Ozone layer depletion—Juvenile literature.   2. Ozone layer depletion—Environmental aspects—Juvenile literature.   [1. Ozone layer depletion.]   I. Title.
II. Series.   QC879.7.P75 1995   363.73'84—dc20   94-25928   CIP AC

This book is printed on 100-percent recycled paper.

CONTENTS

INTRODUCTION

Miles over our heads, a thin layer of ozone gas blankets the earth. It blocks powerful ultraviolet light from reaching the earth and makes life possible on the planet. An international network of chemists and atmospheric scientists has found that this ozone layer has grown thinner over the entire world, and especially over Antarctica. Their studies have identified the cause of the ozone loss: chemicals made and used by people and released into our atmosphere.

Do humans face an ozone crisis, as some claim? Or is the whole matter probably nothing to worry about, as others maintain?

Vanishing Ozone answers these questions. It begins with an explanation of the ozone molecule and the structure of the atmosphere, then traces the fascinating trail of evidence uncovered by scientists all over the world to show that the ozone layer is indeed thinning. It tells why and how economic and political forces have resisted changes that many scientists believe are urgently needed.

Finally, this book describes the scientific questions that remain, and the questions that every person must answer after weighing the evidence.

1

THE JEKYLL-HYDE MOLECULE

In his novel *Strange Case of Dr. Jekyll and Mr. Hyde,* Robert Louis Stevenson wrote of two characters who shared a home in London. One was Dr. Henry Jekyll, a tall, handsome gentleman of good reputation. The other was Mr. Edward Hyde, a short, repulsive-looking man with murderous impulses. Near the end of the story the reader learns that they are the same person, transformed by a chemical potion.

The names Jekyll and Hyde have since been applied to the good and evil in many people, things, and situations. In the past three decades an uncommon blue-colored gas in the atmosphere, ozone, has been called the Jekyll-Hyde molecule. In the lower atmosphere it is Mr. Hyde, harmful to plant and animal life, including humans. High overhead in the upper atmosphere, ozone is Dr. Jekyll, protector of all living things. Yet chemically it is the same molecule.

Ozone is produced when solar energy reacts with pollutants from factories, power plants, and vehicles.

Ozone is a form of oxygen. An oxygen molecule consists of two oxygen atoms, an ozone molecule of three. For as long as the earth's atmosphere has existed, small amounts of ozone have been produced by natural processes. Some ozone molecules form when lightning passes through air. The electricity breaks oxygen molecules apart, freeing oxygen atoms that then form bonds with other oxygen molecules, yielding ozone. You may have smelled the metallic odor of ozone during a thunderstorm. (Early scientists also noticed that same sharp smell after a lightning strike. In 1840 a German chemist, Christian Schönbein, named this gas ozone, based on the Greek word *ozein*, "to smell.") You may also smell ozone wherever electricity is discharged into the air—near a power line, for instance, or even near a toy electric train.

Ozone is also produced when sunlight reacts with hydrocarbons, gases that are given off by trees and other plants, and nitrogen oxides, gases that are given off by microbes in the soil. The amounts of

During a thunderstorm, lightning smashes oxygen molecules apart, creating the metallic-smelling gas, ozone.

ozone produced by natural processes are usually too small to harm living things, but modern life generates much more. Factories, refineries, dry-cleaning plants, and coal- and oil-burning power plants, as well as automobiles and trucks, spew out hydrocarbons and nitrogen oxides in their waste gases. When sunlight acts upon this chemical brew of pollutants, large amounts of ozone are produced.

As you might expect, the highest levels of ozone are found in urban areas, but ozone is carried by winds into rural areas, even into wilderness. Harmful levels of ozone have been measured in national parks. In 1992 a vast cloud of ozone—two thousand miles across and six miles deep—was discovered over the south Atlantic Ocean. It was a result of the tropical sun acting upon waste gases produced by forest and grassland fires in Brazil and other South American countries.

In humans and other animals, ozone affects the respiratory system. It irritates lung tissues and constricts them, reducing lung capacity. It causes chest pains, coughs, headaches, and drowsiness. People exposed to ozone report that they have trouble concentrating on their work. People working or exercising outdoors suffer more than those inside, and those suffering from asthma and emphysema are especially vulnerable to ozone. Medical researchers have found a correlation between rising ozone levels and the number of people hospitalized with asthma attacks. Children are more vulnerable than adults to harm from ozone; they inhale much more air for their size than adults, and they also tend to spend more time outdoors.

Ozone is the most dangerous pollutant in smog. On many summer days, harmful levels of ozone are inhaled by the populations of a hundred cities across the United States. Los Angeles has by far the worst ozone problem in the United States; its air-quality-management district issues more than sixty smog alerts each summer. During a smog alert, schools are urged to reschedule athletic practices and games, and people who are particularly vulnerable to ozone are advised to stay indoors.

It takes very little of this Mr. Hyde molecule to do harm. The United States Environmental Protection Agency (EPA) has set .12 parts of ozone per million parts of air as the health standard. However, scientists in both Canada and the United

Children with asthma are especially vulnerable to harm from ozone pollution.

States have found that ozone levels as low as .08 parts per million impair the lungs of healthy men who are exercising. Environmental groups and the American Lung Association have urged the EPA to set a tougher health standard for ozone.

In addition to affecting the health of people, livestock, pets, and other animals, ozone damages trees, food crops, and other plants. It stunts growth and reduces flowering—and the foods that develop from flowers. The U.S. Department of Agriculture estimates that ozone causes more than a billion dollars of damage each year to timber and to vegetable, fruit, and grain plants. Ozone affects other materials as well; for example, it causes the rubber in automobile tires to deteriorate.

Clearly, ozone in the lower atmosphere has earned its Mr. Hyde label. But in the stratosphere it is Dr. Jekyll. And this side of ozone confuses some people. How can such a harmful chemical also be beneficial to life? they wonder. Also, why don't the increasing amounts of ozone close to the earth's surface rise and replenish some of the ozone that is being lost far above? The answers to both questions lie in the structure of the atmosphere itself.

Most of what we know about the earth's atmosphere has been learned in the past century. In the late nineteenth century scientists sent kites and balloons aloft with instruments that recorded temperatures and gases. From them they learned that the temperature and density of the air decreased with altitude. (Today we know that at 33,000 feet both air pressure and oxygen levels are about a third of their amounts at sea level.) They also found that air temperature fell a degree for every 360 feet of altitude. So they calculated that the temperature at about 140,659 feet would reach absolute zero, the lowest temperature thought to be possible. Beyond that, they reasoned, was outer space.

In 1902 a French meteorologist, Léon Teisserenc de Bort, challenged this idea. Over a span of several years he had sent aloft more than two hundred balloons equipped with devices to measure and record the temperature as the balloon rose. At 25,000 feet—nearly five miles up—his instruments showed that temperatures stopped decreasing. In fact, the air began to grow warmer.

De Bort's findings were confirmed by other meteorologists. In 1908 de Bort named the lower atmosphere the troposphere, after the Greek word *trepein* ("to turn"), because of its rising and falling air currents. For the mysterious region above that he suggested the name stratosphere (from the Latin *strātum*, "a covering"), because it seemed to have no air movements. The boundary area in between, where air temperatures begin to warm, was later named the tropopause.

The relatively warm layer of the tropopause is not a perfect seal between the troposphere and stratosphere. There is some interchange of gases and tiny particles between these two regions, but the ozone of the lower atmosphere never reaches the stratosphere. This is because ozone is a rather unstable molecule. It readily gives up one of its oxygen atoms to other gases and doesn't last long enough to rise to the stratosphere.

The troposphere extends upward an average of seven and a half miles but is highest over the equator and lowest at the poles. It also changes with the seasons, expanding upward over the half of the earth that is experiencing summer and shrinking in height over the winter half.

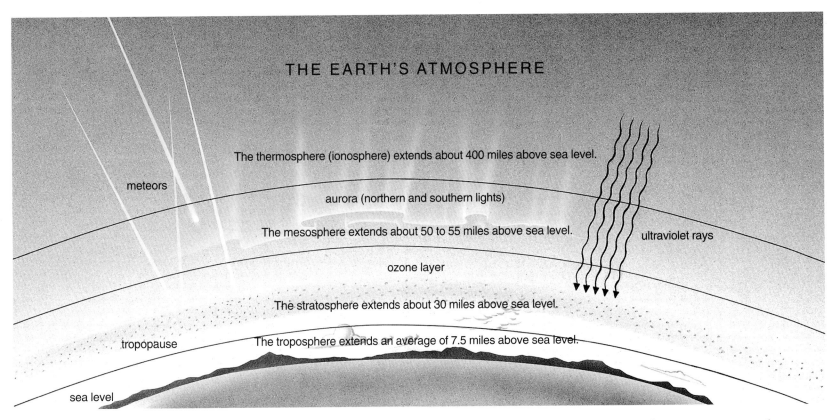

THE EARTH'S ATMOSPHERE

meteors

The thermosphere (ionosphere) extends about 400 miles above sea level.

aurora (northern and southern lights)

The mesosphere extends about 50 to 55 miles above sea level.

ultraviolet rays

ozone layer

The stratosphere extends about 30 miles above sea level.

tropopause

The troposphere extends an average of 7.5 miles above sea level.

sea level

In the late nineteenth century, even before the stratosphere was named, scientists began to suspect that ozone existed up there. Something was preventing the full range of the sun's ultraviolet light from reaching the earth's surface. A British chemist, W. N. Hartley, conducted laboratory experiments with ozone and ultraviolet light. He concluded that ozone was the most likely suspect, and that it must be more abundant in the upper atmosphere than near the ground. (In fact, about 90 percent of all ozone is in the stratosphere.)

By the 1920s new types of measuring devices had helped scientists prove the existence of an ozone layer in the stratosphere. And in 1931 Sydney Chapman, a British geophysicist, offered an explanation of how ozone forms in the stratosphere. He said that in this region oxygen molecules are broken up by the shortest and most powerful ultraviolet rays, UV-C. This frees oxygen atoms, which then bond with other oxygen molecules to form ozone molecules. Other strong ultraviolet rays, called UV-B, cause ozone molecules to split, freeing oxygen atoms and creating oxygen molecules. Normally these oxygen-to-ozone and ozone-to-oxygen reactions balance each other, so the amount of the stratosphere's ozone remains about the same.

Scientists today know that stratospheric chemistry is much more complex than Chapman proposed, but he was basically right, and he also pointed out the importance of the ozone layer to life on earth. When UV-C and UV-B radiation break up oxygen or ozone molecules, heat energy is given off. This is the heat that warms the tropopause and the stratosphere. Loss of the ozone layer would have huge effects on the earth's climate.

The ozone layer also keeps all UV-C radiation from reaching the earth's surface—fortunately, because this most powerful ultraviolet light is a death ray. Ozone also blocks most UV-B radiation. The strongest UV-B rays that do reach the earth's surface can cause skin cancers and cataracts in people. They damage crops and microorganisms that are the foundation of many food chains. This damage is limited—so far—because the ozone layer prevents nearly all of this powerful ultraviolet light from reaching the earth's surface.

Ozone is just a trace gas in the stratosphere, spread thinly through a vast area. Only about eight in one million molecules are ozone molecules. If all of the stratosphere's ozone was brought down to the earth's surface, the air pressure here would compress it into a layer just one eighth of an inch thick. Looked at that way, the earth's ozone layer seems very slight, a fragile defense against deadly rays from the sun. In 1933 Dr. Charles Abbot, of the Smithsonian Institution in Washington, D.C., said, "It is astonishing and even terrifying to contemplate this narrow margin of safety on which our lives thus depend. Were this trifling quantity of atmospheric ozone removed, we should all perish."

How ironic that humans, the only creatures on earth intelligent enough to figure out the vital importance of the ozone layer, have for decades released chemicals that have been eating it away.

2

DISCOVERY OF THE OZONE THREAT

When you look up at a clear blue sky, the atmosphere may seem simple, but this membrane of air surrounding the earth is a complex thing. The air's movements and chemistry are affected by the sun and by the planet itself, including the earth's land surface, oceans, plants, and animals. The more scientists learn about the atmosphere, the more complicated they realize it is.

Scientists have learned more about the chemistry and dynamics of the atmosphere in the past quarter century than in all the centuries before. New devices and techniques have helped make this possible. However, the main reason for this great increase in knowledge is concern about human-made threats to the atmosphere.

This worry has inspired a huge research effort involving thousands of scientists from numerous countries. Their findings have been presented at conferences

Viewed from space, the cloud patterns over the earth
demonstrate some of the complexity of its atmosphere.

and published in dozens of books and scores of scientific journals. This book does not try to explain every facet of this search for understanding. (If it did, the book might be too large and heavy for you to hold!) Instead, it introduces some of the key scientists and describes some of the major discoveries. It omits a lot of the chemical complexity but explains how scientists have reached our present understanding of the threatened ozone layer. Like all environmental problems, this one involves more than science and technology; it also involves economic and political factors, and the response of powerful businesses when their products or processes are found to be harmful. These factors, too, are important.

The first burst of atmospheric research in modern times occurred in the late 1960s. It was sparked by the development of the supersonic transport (SST), a passenger aircraft that was designed to cruise in the stratosphere. The SST would travel faster than the speed of sound, thereby reducing long flights to a few hours. Its advocates imagined a future when thousands of SSTs would fly routes worldwide every day.

The first environmental concerns about the SST focused on sonic booms—the loud sounds of shock waves created when a supersonic aircraft compresses the air as it flies. By 1967 there was growing public opposition to United States development of an SST fleet. While scientific studies of sonic-boom effects drew lots of attention, a few scientists were investigating the possible effects of water vapor and exhaust gases that SSTs would release into the stratosphere.

Some studies showed that SST exhaust gases would cause a decrease in global ozone. In 1971 Dr. James McDonald, an atmospheric physicist at the University of Arizona, testified before a congressional subcommittee that flights by hundreds of SSTs would destroy ozone, thereby allowing more ultraviolet radiation to reach the earth's surface and causing an increase in human skin cancers. His testimony was attacked by SST advocates, but this new concern was another nail in the coffin of a U.S.-built SST. Later in 1971 both the U.S. Congress and Senate voted against developing an American supersonic transport. (Today a small fleet of Concorde SSTs is operated by Air France and British Airways. Because they are few in number and because they fly low in the stratosphere, these aircraft are not considered a serious threat to the ozone layer.)

The SST controversy focused attention, and government research funds, on the upper atmosphere and the worrisome possibility that human technology could affect the earth's ozone shield. In 1972 James Lovelock, a British scientist who had invented a highly sensitive device for measuring tiny amounts of gases, discovered traces of two types of chlorofluorocarbons (CFCs) in the atmosphere. These are human-made chemicals that do not occur in nature.

The usefulness of CFC gases had begun to be recognized by chemists and manufacturers in the early 1930s, when CFCs were first used as refrigerants. Coolant gases in refrigerators must be nonpoisonous, nonflammable, and stable. CFCs have these qualities. They are inert: They do not normally react with other chemicals. They were perfect for this job—and others.

By the early 1970s CFCs were the most common coolants in both refrigerators and air conditioners. They were also widely used as expansion agents, changing plastic pellets into plastic foam products, and as cleaning agents in the manufacture of electronic parts, including computer chips. Furthermore, CFCs proved to be

Until the late 1980s, CFCs were sometimes used in the manufacture of packing materials like these and other plastic foam products.

19

ideal propellant gases in spray cans. They helped produce a fine spray of droplets (an aerosol) and did not react with the liquid product in the cans. CFCs were used in hundreds of spray-can products.

As a result of these and other uses, almost a million tons of CFCs were released into the atmosphere each year.

An American scientist, Dr. Sherwood "Sherry" Rowland, a professor of chemistry at the University of California at Irvine, wondered what became of these chemicals after they were released into the atmosphere. In 1973 he encouraged Dr. Mario Molina, a postdoctoral student in his department, to investigate.

Molina first tried to find some way in which CFCs would break down in the lower atmosphere. He found none. This meant that CFC molecules would slowly rise into the stratosphere. There ultraviolet light would break them down. (This had been learned in previous studies by Canadian scientists.) The chemical reaction would free some chlorine atoms. Before Molina published the results of his studies, Rowland suggested that he answer one further question: What happens to the chlorine?

The research of Dr. Sherwood Rowland (left) and Dr. Mario Molina (right) showed that chlorine freed from CFCs would harm the earth's ozone layer.

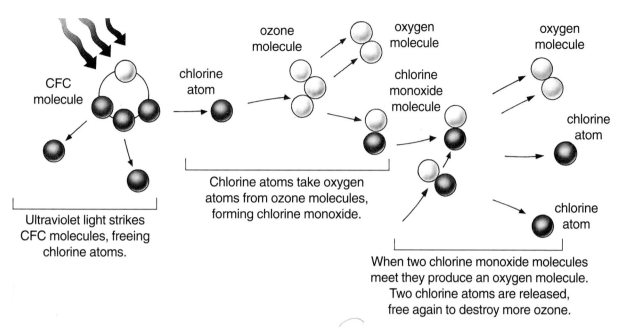

Ultraviolet light strikes CFC molecules, freeing chlorine atoms.

Chlorine atoms take oxygen atoms from ozone molecules, forming chlorine monoxide.

When two chlorine monoxide molecules meet they produce an oxygen molecule. Two chlorine atoms are released, free again to destroy more ozone.

How CFCs Destroy Ozone

Molina's calculations revealed that just one free chlorine atom could, through a chain of reactions, destroy tens of thousands of ozone molecules. This "chlorine chain" was a startling discovery. Molina was alarmed. He thought he might have made a mistake. He and Rowland carefully rechecked his calculations. The results were the same: Chlorine freed from CFCs could destroy ozone.

The damage to the earth's ozone shield would depend on how many CFCs reached the stratosphere. In the early 1970s more and more of these molecules were released each year, and they would continue to drift slowly upward. Each CFC molecule takes between six and eight years to reach the stratosphere. Using CFC production figures from 1972, Rowland and Molina calculated that as much as 20 to 40 percent of the ozone layer might be depleted within a century.

These findings were published in the June 28, 1974, issue of *Nature,* a prestigious British scientific journal. The report on atmospheric chemistry included a call for further studies, "in order to ascertain the levels of possible onset of environmental problems." Although scientific interest in the ozone-depletion theory began to spread around the world, the news media paid no attention. Rowland and

Molina realized they had to take further steps in order to alert the public. The threat of vanishing ozone was too serious. They felt, Molina recalls, that "we had a *responsibility* to go public."

Their opportunity came at the 1974 convention of the American Chemical Society. Rowland and Molina prepared a 150-page report on their theory. (In casual conversation the word *theory* often means an untested idea. In science an untested idea is a hypothesis. A theory is supported by solid evidence.) The chemical society's news manager recognized the importance of their report and scheduled a press conference for Rowland, who presented updated calculations: CFCs would cause a 5 to 7 percent loss of ozone by 1995 and a 30 to 50 percent loss by the year 2050. He warned of possible changes in the earth's climatic patterns and of increased skin cancers. Rowland estimated that a 5 percent decline in stratospheric ozone would cause forty thousand extra cases of skin cancer each year in the United States alone.

The press conference inspired some sensational headlines—AEROSOL SPRAY CANS MAY HOLD DOOMSDAY THREAT—but also drew the attention of other scientists, who began to report evidence that supported the ozone-depletion theory. Public concern grew. Environmental groups and scientists called for an end to CFC production.

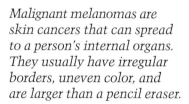

Malignant melanomas are skin cancers that can spread to a person's internal organs. They usually have irregular borders, uneven color, and are larger than a pencil eraser.

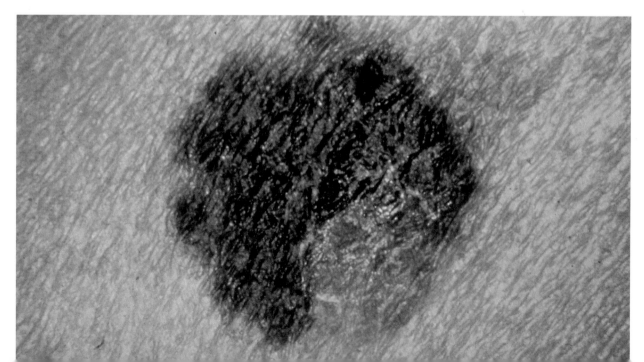

The autumn of 1974 marked the beginning of a conflict that has been called the Ozone War. On one side were chemists and atmospheric scientists from all over the world, plus environmental and consumer groups and some political leaders. They believed the threat to the ozone layer was real and that CFC production should be halted. On the other side were several powerful industries (including CFC, spray-can, and refrigeration interests) and their supporters, which included conservative politicians and columnists who opposed government regulation of business.

In the mid-1970s, chlorofluorocarbons were the basis of a thriving $8 billion industry that provided more than two hundred thousand jobs just in the United States. One major chemical company, Du Pont, had financed research on CFCs that showed that these molecules caused no harm in the lower atmosphere. The news that they were destroying the stratosphere's ozone shield was a terrible blow. Investments, profits, and careers were threatened. The CFC industry and its supporters responded by attacking the scientific evidence of the ozone-depletion theory and by arguing that much more study was needed on the "alleged" threat.

The first target of consumer and environmental groups in the Ozone War was the use of CFCs in spray cans. Despite industry claims, substitutes for CFCs in spray cans were available. Furthermore, although CFCs escaped gradually via leaks from such products as refrigerators and air conditioners, they were released immediately into the atmosphere every time a spray can was used. Spray cans also accounted for more than half of all the CFCs used in the United States. A ban on the use of CFCs in spray cans would be a giant positive step, especially if other nations followed.

As it turned out, Sweden was the first nation to ban CFC propellants, in 1976. A year earlier, the state of Oregon had done the same (with the ban taking effect in early 1977). However, a national ban in the United States was stalled by the powerful forces opposed to change. The chemical industry spent millions of dollars defending CFCs and attacking the ozone-depletion theory. For example, the business

Spray cans were once a major source of CFC pollution in the United States. Substitute gases are available, and nearly all use of CFCs in spray cans stopped in the late 1970s.

organization now called the Chemical Manufacturers Association sponsored scientists who questioned the theory at press conferences and at government hearings. One such "expert" was a British professor who was an authority on clouds but not on atmospheric chemistry. He was hired for a U.S. speaking tour during which he denounced the ozone-depletion theory as "utter nonsense."

Industry spokespeople fostered the notion that scientists disagreed about whether CFCs harmed earth's ozone shield. Although sales of spray cans dropped, many people were led to believe that there was not enough evidence to ban CFCs. In 1975 the "insufficient evidence" argument was used by the U.S. Consumer Product Safety Commission when it refused to outlaw the use of CFCs in certain products, including refrigerators. The commission was responding to a lawsuit brought by the Natural Resources Defense Council, a leading environmental group on the CFC issue.

"Doomsayers," "un-American," and "saboteurs" were some of the labels given to environmentalists and scientists who sought to have CFCs banned. Conservative columnists and industry publicists linked them with Henny Penny, the children's-story character who leaped to the conclusion that the sky was falling when her head was struck by an acorn.

However, independent scientists continued to find evidence that supported the ozone-depletion theory. High-altitude air samples confirmed that CFCs were reaching the stratosphere and being broken down at the rates that Rowland and Molina had predicted. The industry response was that there was no evidence, aside from laboratory studies, that ozone was actually being depleted in the stratosphere.

The opposing forces in the Ozone War awaited the results of a yearlong study by the National Academy of Sciences, which were due in April 1976. Then, early that year, two scientists announced some surprising evidence that cast doubt on the Rowland-Molina ozone-depletion theory. The scientists were Rowland and Molina themselves!

Both Rowland and Molina had carefully searched through decades of scientific literature, looking for evidence that would confirm or refute their theory. They looked for chemicals that would tie up free chlorine and keep it from destroying ozone. One molecule they considered was chlorine nitrate. At first they had dismissed it because German scientists had found that it broke down quickly in sunlight, but they decided to investigate further. With the help of an associate, John Spencer, Molina conducted laboratory tests that revealed that this chlorine compound could last for hours, not minutes, in sunlight. This changed the results of their ozone-loss calculations. It seemed that ozone destruction would be less severe than they had thought. In fact, some calculations suggested that the presence of chlorine nitrate might actually *increase* the normal levels of ozone in the lower stratosphere.

Rowland informed the scientists who were studying ozone depletion for the National Academy of Sciences. Their report was nearly complete when the chlorine nitrate bombshell struck. The April deadline for the report passed as committee members and others investigated the possible effects of chlorine nitrate, weighing the new evidence and its implications.

To some ardent CFC advocates the implications seemed simple and straightforward: The ozone-depletion theory was wrong, and it was back to business with chlorofluorocarbons. But other observers, including some industry scientists, knew that was not the case. After a five-month delay, during which the potential effects of chlorine nitrate were thoroughly explored and found to be less important than when first estimated, the National Academy of Sciences issued two reports in September 1976.

The academy's Panel on Atmospheric Chemistry concluded that CFCs were harming the earth's ozone shield, and this in turn would cause increased skin cancers and climate changes, including a warming of the earth's atmosphere. The report concluded that production of CFCs should be restricted.

Given these findings, the academy's second report, by the Committee on Impacts of Stratospheric Change, was surprisingly timid. It stated that there should be two more years of study and assessment before the federal government acted on the CFC issue.

The academy's scientific conclusions were strong enough, however, to push government agencies into action. By the spring of 1977 the Environmental Protection Agency, the Food and Drug Administration, and the Consumer Product Safety Commission announced that all nonessential uses of CFCs had to be phased out by early 1979.

As manufacturers began using substitutes for CFCs they advertised that their spray cans were "ozone safe" and boasted of their concern for the human environment. To the general public it seemed that the Ozone War was over, but in reality just one battle had concluded in a few countries. CFCs were still used in spray cans in some nations, and most other uses of chlorofluorocarbons continued—and grew.

In 1979 the National Academy of Sciences issued another report on the threat to ozone. It warned again of increased cancer deaths as well as damage to farm crops and ocean food chains. The academy urged the United States to work toward a global ban of CFCs. The CFC industry used this to its advantage, arguing against tougher regulations in the United States until the time when CFC production was reduced all over the world. Industry representatives had monitored international discussions on CFC control; they believed that global action against CFCs lay far in the future.

With the election of President Ronald Reagan in 1980, most progress in environmental regulation in the United States slowed or came to a halt, including efforts to influence other countries to reduce the use of CFCs. One of the president's basic goals was to "get government off the back of business." His administration maintained that U.S. reductions of CFCs had to be tied to international controls—which the United States began to oppose.

Some officials appointed by Reagan began to question the evidence behind the ozone-depletion theory. CFC manufacturers took heart; one company announced plans to expand production. In the early 1980s the Ozone War was far from over, and the tide of battle had shifted.

If CFCs were not kept from the earth's atmosphere, warned the National Academy of Sciences, such basic and vital food crops as wheat might be harmed.

3

DISCOVERY OF THE OZONE HOLE

Boring is a word that can easily be applied to long-term studies that measure and record basic information about the environment. Time and again, however, such records reveal important changes that might have gone undetected. Sometimes they yield startling news.

This was the case with measurements of trace gases in the atmosphere that had been taken every year since 1957 by the British Antarctic Survey. Each year geophysicist Dr. Joseph Farman sent assistants to the British research station at Halley Bay, Antarctica. They used devices called Dobson spectrophotometers to measure the amounts of ozone and other gases in the atmosphere over Halley Bay. Dobson spectrophotometers measure ozone in Dobson units. (Today several dozen of these ground-based devices all over the earth continue to monitor ozone levels.)

Long-term studies of the atmosphere over the vast, frozen
continent of Antarctica led to a startling discovery in 1981.

The worth of this effort had been questioned by Farman's superiors, and several times funds for the long-term monitoring had almost been cut off. Farman managed to keep the program alive. In 1981 his belief in the value of this environmental monitoring was vindicated. The Dobson readings of ozone in September and October—spring in Antarctica—were 20 percent below normal.

Could something be wrong with the equipment? By the next Antarctic spring, Farman had a new Dobson spectrophotometer in place. It too recorded below-normal ozone in the stratosphere over Antarctica. The following spring Farman took Dobson measurements from the Halley Bay station and from an island a thousand miles northwest. The instruments measured a huge drop in ozone over both sites. For many years ozone levels had been around 300 Dobson units. In 1984 the measurement was only 140.

Joseph Farman was well aware of the ozone-depletion theory, the criticism aimed at it, and the strong resistance by industry to reducing CFCs. His own government opposed regulation of CFCs. It would take courage to publish the results of the British Antarctic Survey's ozone measurements, especially since they showed such dramatic change: Almost all computer models based on the ozone-depletion theory predicted that the loss of the earth's ozone shield would be slow and gradual.

Farman had another concern. Since 1978 a satellite launched by the U.S. National Aeronautics and Space Administration (NASA) had been monitoring ozone levels in the stratosphere. The *Nimbus 7* satellite carried ozone-measuring devices that were more accurate than Dobson spectrophotometers, yet NASA had reported nothing amiss.

Still, Dr. Farman believed in the strength of the data gathered by the British Antarctic Survey. With help from three colleagues he wrote a report that was published in the May 16, 1985, issue of *Nature*. The article described the startling change in ozone levels and pointed out that the drop in ozone coincided with a rise in CFCs over Antarctica.

The journal article aroused interest among atmospheric scientists all over the

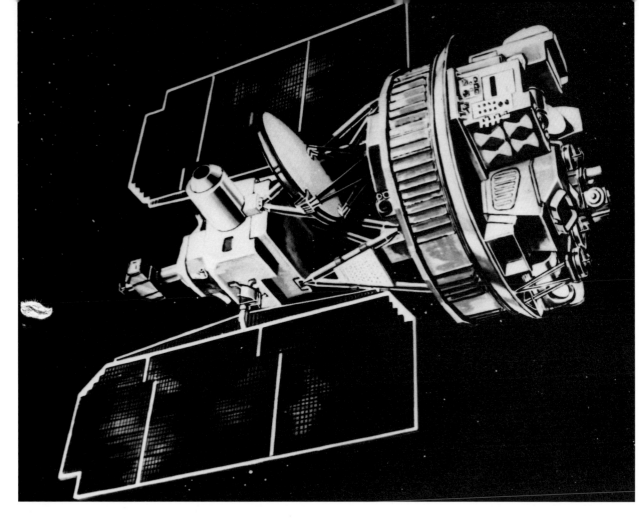

Instruments aboard the Nimbus 7 satellite confirmed the annual spring loss of ozone over the Antarctic region.

world, and especially at NASA. Researchers there wondered why the *Nimbus* 7 instruments had detected no ozone loss. They discovered that the loss *had* been measured and recorded, but not reported. The satellite's instruments poured out so much data that its computers had been programmed to report only ozone readings in the normal range. They were to ignore readings below 180 Dobson units as "impossible." When the full range of measurements was checked, NASA's figures confirmed the British observations.

Data from *Nimbus* 7 also revealed the size of the area over Antarctica where ozone dropped sharply each spring. It was the size of the continental United States. This area is called the ozone hole, although it is more accurately a seasonal thinning of the ozone layer, and therefore a weak place in earth's defense against ultraviolet radiation.

In the early 1980s the worldwide production of CFCs grew despite efforts by environmental groups and warnings by such scientists as Sherry Rowland. In 1984, while Joseph Farman and his colleagues waited for another Antarctic spring in order to confirm their worrisome findings, Rowland said, "From what I've seen over the past ten years, nothing will be done about this problem until there is further evidence that a significant loss of ozone has occurred. Unfortunately, this means that if there is a disaster in the making in the stratosphere, we are probably not going to avoid it."

The ozone hole was certainly evidence of a significant loss of ozone, and its

Data from satellites are converted to computer-made images that show the amount of ozone over Antarctica in different seasons and years.

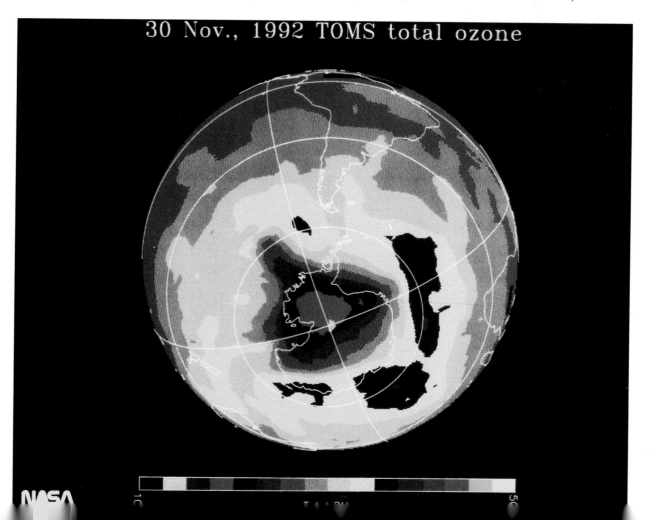

30 Nov., 1992 TOMS total ozone

NASA

discovery turned the tide of battle in the Ozone War. In 1986 Rowland stated, "Industry always said that we'd have plenty of advance warning of any ozone problems, but now we've got a hole in our atmosphere that you could see from Mars."

There was also growing concern about global warming—the threat of global climate change caused by pollutants in the atmosphere. Although increasing amounts of carbon dioxide, methane, and other gases are the main cause of global warming, CFCs also add to the problem. While scientists all over the world studied the cause of the ozone hole and its significance, steps were taken toward reducing CFC production.

In 1985 twenty nations met in Vienna, Austria, and signed an agreement that recognized the problem of ozone loss. The document called for international cooperation in sharing the results of research and set some goals in ozone and chlorofluorocarbon studies. Negotiations among nations continued, eventually leading to a September 1987 meeting in Montreal, Canada. Representatives of forty-three countries attended, with the goal of cutting global CFC production. The United States delegation, representing the Reagan administration, sought to weaken the effort in its attempts to protect the U.S. CFC industry. Eventually, however, the agreement was signed by all forty-three countries present. The Montreal Protocol on Substances That Deplete the Ozone Layer set target dates for reducing CFC production, with output of CFCs to be cut in half by the end of 1999.

In the opinion of environmental groups and many scientists, the Montreal Protocol was a small step in a situation that demanded big strides. Nevertheless, it did call for nations to meet regularly to respond to new scientific evidence. Atmospheric scientists agreed on one basic point: that free chlorine from CFCs caused the spring ozone hole over Antarctica. But many questions remained: Why did the great loss of ozone occur there? And why during the spring?

The Antarctic continent and the air above it became the focus of research. One scientist who led a 1986 expedition to Antarctica was Dr. Susan Solomon, an atmospheric chemist with the National Oceanic and Atmospheric Administration

Atmospheric chemist Dr. Susan Solomon led the 1986 expedition to investigate the ozone "hole" above Antarctica.

(NOAA) in Boulder, Colorado. In early 1985 she had been one of the experts asked by the scientific journal *Nature* to review Joseph Farman's report to see if it was worthy of publication. In a later interview she said, "The idea was so incredible, I think my first response was to wonder, 'Who are these people, are they crazy?' But what was so convincing was the good job they had done.... They had years of good data and measurements from two ground locations."

Solomon was an expert on both the chemistry and the dynamics—air movements—of the atmosphere. With another scientist, Rolando Garcia, she tried to duplicate the large spring ozone loss on a computer model but failed. She conferred with Sherry Rowland, and they both began to focus on the possibility that something unusual in the atmosphere over Antarctica caused the sudden loss of ozone there. Then, in late 1985, Solomon attended a meeting of the American Geophysical Union and heard a report, given by Dr. David Hofmann of the University of Wyoming, on unusual clouds that occur in the Antarctic stratosphere.

Polar stratospheric clouds were observed by early explorers of both the Arctic and Antarctic. These clouds are especially common over Antarctica because of its extreme cold. In the winter, air over the Antarctic region circulates in a pattern called the polar vortex; it goes around and around without any inflow of air from warmer regions. Polar stratospheric clouds occur when air temperatures drop below –112 degrees Fahrenheit (–80 degrees Celsius). The clouds are made of ice particles containing water and nitrogen compounds.

These clouds exist all through the long Antarctic winter. David Hofmann had studied them with devices carried aloft by balloons. One of the charts used to illustrate his findings showed the vertical location of the clouds. They were most dense

Polar stratospheric clouds play a key role in the springtime loss of ozone over the Antarctic and Arctic.

between six and fifteen miles above the earth. Susan Solomon recognized that this coincided closely with the altitude of greatest ozone loss, as reported by *Nimbus 7*. She called Rowland about this new evidence.

After further study, Solomon, Rowland, Garcia, and a fourth scientist, Don Wuebbles, coauthored a report, "On the Depletion of Antarctic Ozone," in the June 19, 1986, edition of *Nature*. They argued that polar stratospheric clouds were a key factor in the great loss of ozone over Antarctica. The surfaces of the ice crystals making up the clouds served as a storehouse of chlorine compounds during the winter. Then spring sunshine melted the cloud ice crystals, releasing huge amounts of chlorine that quickly depleted the ozone layer. The ozone was replenished during the Antarctic summer, when warm air broke up the polar vortex and also brought ozone from warmer areas of the Southern Hemisphere.

This chemical explanation of the ozone hole was hailed by some atmos-

In Antarctica and elsewhere on earth, scientists use large balloons to carry air-sampling equipment and other devices high in the atmosphere.

pheric scientists but rejected by others. A few scientists believed that evidence pointed to *natural* causes. One theory was that radiation from storms on the sun, called sunspots, which occur in cycles, triggered nitrogen reactions that destroyed ozone. This was called the solar theory. Another explanation was that air movements in the stratosphere carried ozone-poor air to the Antarctic region and made it appear that ozone had been destroyed. This was called the dynamics theory.

Scientists all over the world were eager to get more data from Antarctica. A good time for vital observations was August—winter in Antarctica—when the chemistry of the stratospheric clouds could be studied directly. In August 1986 a team of scientists led by Susan Solomon arrived in Antarctica. This first National Ozone Expedition (NOZE) used instruments on the ground and others carried aloft by balloons. By mid-October the team of scientists had evidence that discounted the solar and dynamics theories and supported the chemical theory.

Even as the first NOZE team was at work, a second expedition was being planned for the Antarctic winter of 1987. It used two aircraft, including an ER-2 that was specially designed for high-altitude flight. The ER-2 collected data about stratospheric chemistry as it flew through the ozone layer at altitudes above sixty

Originally designed as a high-altitude photo spy plane, the ER-2 proved useful for learning about chemical reactions in the stratosphere.

thousand feet. The results confirmed and strengthened the finding of the first NOZE expedition: Chlorine from human-made CFCs was destroying the ozone layer over more than twelve million square miles of Antarctic land and sea.

Even in the face of mounting evidence, the CFC industry fought on. In early 1988 it found ammunition in an article in the journal *Science* that reported that harmful UV-B radiation was actually decreasing in the United States. Measurements taken near eight U.S. cities between 1974 and 1985 recorded small declines in UV-B, not the increase that one might expect if the ozone layer was being depleted. The CFC industry hailed the report as evidence that their product was harmless. However, atmospheric chemists had another explanation for the decrease in UV-B radiation near cities: In urban areas a sort of pollution shield, including ground-level ozone, was helping the stratospheric ozone shield to block some ultraviolet rays.

Ever since the discovery of the ozone hole over Antarctica, scientists had looked for evidence of ozone thinning elsewhere. The Ozone Trends Panel, an international group of 130 scientists whose work was coordinated by NASA, began to investigate this matter in 1986. The scientists analyzed data from all over the world, and from both Dobson spectrophotometers and ozone-measuring devices on satellites called total ozone mapping spectrometers (TOMS).

The panel announced its findings in March 1988. It reported losses of ozone all over the world, including the Northern Hemisphere, between 1978 and 1985. The ozone shield was thinning over China, Japan, Europe, what was then called the Soviet Union, Canada, and the United States. The losses, up to 6 percent a year, were greatest in the wintertime but occurred in all seasons. The panel's report stated that more ultraviolet light was reaching the earth than at any time in modern history. Pollution was probably blocking some of it, but the increased UV-B radiation would cause more skin cancers and other harm.

Just a few days after the report of the Ozone Trends Panel, Du Pont announced that it had accepted the evidence that CFCs were harming the earth's ozone layer.

After fourteen years of resistance, Du Pont, producer of a quarter of the world's CFCs, was getting out of the business as soon as substitute chemicals could be developed. Other manufacturers soon followed.

Although CFC makers and other businesses—for example, refrigerator and air-conditioner manufacturers—had spent millions of dollars in efforts to deny and discredit the scientific case against CFCs, they had also hedged their bets. Since the mid-1980s they had been quietly searching for alternatives to CFCs. (For details about the chemicals that are being substituted for CFCs, see "Good-bye to CFCs," on pages 57–58.)

The nations that had negotiated the 1987 Montreal Protocol met again in 1990 and 1992, each time aiming to phase out CFCs sooner than before. The revised protocol also called for a ban on halons, which are ozone-destroying gases used in fire extinguishers. Poorer countries were allowed more time to stop using CFCs, but wealthier nations promised to help these countries switch to alternatives. The United States agreed to give at least $40 million to this effort. Furthermore, the United States and the European Community of nations planned to cease nearly all CFC production by the end of 1995.

At last, after years of political struggle, big strides were being taken to protect the earth's ozone layer. In 1993 Dr. James Elkins of NOAA reported that the buildup of CFCs in the atmosphere was slowing. The growth rate of one type, CFC-11, had averaged eleven parts per trillion from 1985 to 1988. In 1993 it was three parts per trillion. The growth rate of another kind, CFC-12, also dropped. It had averaged twenty parts per trillion between 1985 and 1988. In 1993 it was eleven parts per trillion.

These encouraging measurements did not signal that the threat was over. Each CFC molecule has a long lifetime—about forty years for CFC-11 and more than a hundred years for CFC-12. In addition, despite laws in the United States that call for recycling CFCs from old appliances, some will continue to escape. In the United States alone there are 160 million refrigerators, 30 million freezers, 45 million home

air conditioners, and 90 million auto air conditioners that contain chlorofluorocarbons. CFCs may arise from other sources, too, including plastic-foam packaging buried in landfills.

Atmospheric scientists and environmentalists applauded the phasing out of ozone-destroying chemicals, but their pleasure was muted by scientific reality. There were already enough CFCs in the atmosphere to cause further thinning of the ozone layer. During the last years of the twentieth century and for the first decade or so of the twenty-first, life on earth will be exposed to more and more harmful ultraviolet rays. Things will get worse before they get better.

CFCs are now recovered from many discarded refrigerators but can escape into the atmosphere from those that are carelessly disposed of.

4

QUESTIONS FOR SCIENCE
AND FOR SOCIETY

In the Antarctic spring of 1993, the ozone hole was the worst to date. Just 90 Dobson units were measured; the ozone layer was only a third of its normal thickness. And in the Northern Hemisphere, spring 1993 ozone levels were down 10 to 20 percent from their normal range. The same low levels were measured in 1994.

Chemicals from a volcanic eruption may have contributed to the ozone loss. The idea that volcanoes can make ozone depletion worse was first proposed in 1989 by Susan Solomon and David Hofmann. Then, in June 1991, the Philippine volcano Mount Pinatubo exploded and spewed forty billion pounds of sulfur dioxide gas into the upper atmosphere. This vast cloud reflected solar energy back into space and brought a temporary halt to global warming. In fact, it caused a slight cooling of the earth's atmosphere.

The 1991 eruption of Mount Pinatubo had a marked but temporary effect on the atmosphere's temperature and its ozone layer.

However, like ice crystals in polar stratospheric clouds, tiny droplets of acid from the Pinatubo eruption apparently served as surfaces for chemical reactions that destroyed ozone. In 1993 the Pinatubo cloud contributed to ozone declines over most of the world. Over the United States the ozone shield was 10 percent thinner than usual during the summer of 1993. This allowed harmful UV-B rays to increase by at least 20 percent.

Pinatubo was the greatest volcanic eruption of the century, but its effects on the earth's atmosphere were temporary. Most of the debris from Pinatubo settled out of the atmosphere in a few years, so its impact on both atmospheric temperature and the ozone layer was brief. Mount Pinatubo had added another layer of complexity to the mysteries of the earth's atmosphere. Scientists looked forward to the ozone measurements of 1995 and later, when the volcanic effects would be small and changes caused by CFCs would be more clear.

Perhaps "looked forward to" is the wrong expression, because scientists were worried about the ozone losses over the Northern Hemisphere. Consider the Arctic, for example. Although its climate is warmer than the Antarctic's, some polar stratospheric clouds form there as well—and help chlorine eat away the ozone layer. The loss is less there than over Antarctica, but it is nonetheless important, because many more people live in and near the Arctic. They will be exposed if harmful solar radiation increases.

Some atmospheric scientists believe that air movements in the stratosphere play a vital part in determining which areas—and which people—are exposed to harmful ultraviolet radiation. For example, in late winter and spring, ozone-depleted air from the Arctic spreads outward, probably causing the weakened ozone layer that has been measured over Sweden. In 1990 scientists reported that masses of ozone-poor air flowed northward during the 1987 Antarctic spring and created temporary ozone "holes" over heavily populated areas of southern Australia. The amount of hazardous UV-B rays increased by 14 percent in the month of December—the beginning of summer, and the beach season, in the Southern Hemisphere.

Long, unprotected exposure to sunlight ages skin prematurely and can cause disfiguring and even fatal cancers.

Ozone-poor air from Antarctica also spreads out over Chile, Argentina, and New Zealand. The Australian government encourages its citizens to protect themselves with clothing, hats, sunglasses, and sun-blocking lotions. In advertisements it urges them to "Slip, Slop, Slap"—slip on a shirt, slop on sunscreen, and slap on a hat. Hats are a required part of school uniforms, and such outdoor activities as sports are scheduled in the late afternoon, when solar radiation dims. Also, scientists who work in Antarctica itself are warned about the UV-B rays slicing through the thin ozone of the Antarctic spring.

According to a conservative estimate, loss of just 5 percent of the earth's ozone shield would cause an additional five hundred thousand cases of skin cancer each year. The most common kinds of skin cancer are seldom fatal, but melanoma usually is, and the number of melanoma cases has grown dramatically. Like most cancers, melanoma does not develop quickly. It may appear ten or even twenty years after exposure to ultraviolet rays—years after a young person suffers a severe sunburn, for example. (There is abundant evidence that exposure to sunlight causes melanoma cancers, but the specific cause may not be UV-B rays. Some evidence points to UV-A rays, which penetrate deeper into the skin than UV-B rays. Scientists are trying to identify the specific cause.)

Powerful wavelengths of ultraviolet light can also cause eye cataracts and damage the immune system, making the body more vulnerable to disease. But people can protect themselves from the dangers of UV-B rays. (For some hints about this, and other actions to take on the issue of the threatened ozone layer, see "What You Can Do," on pages 55–56.)

However, other living things cannot consciously protect themselves, and ever since the CFC threat to the ozone layer was first identified, scientists have tried to assess the danger of stronger doses of ultraviolet light to plants and animals. In 1994, large numbers of kangaroos in southeastern Australia went blind. Whether the cause was UV light or disease is being investigated.

The wildlife of Antarctica has received special attention. Such animals as penguins and seals are shielded from UV-B rays by feathers or fur, and penguin and seabird eggs are opaque and thus protected from ultraviolet light.

Increased ultraviolet light could harm penguins indirectly by reducing their food supply.

The shrimplike crustaceans called krill—up to 1 1/2 inches long—are vital food for many Antarctic creatures, from small fish to mighty whales.

Powerful UV wavelengths can damage their eyes, though. These animals could also be harmed in another, indirect way: through depletion of their food supply. Environmental groups and biologists have warned that ultraviolet light might harm phytoplankton, the microscopic drifting algae that are the first link in many Antarctic food chains. Small crustaceans called krill are the next link in many food chains, and fish, penguins, seabirds, seals, and certain kinds of whales feed on krill. In 1989 Dr. Sayed El-Sayed, a plant ecologist at Texas A&M University, said, "Krill depend on phytoplankton, and if anything happens to krill, the whole ecosystem will collapse."

For good reason, then, biologists have focused attention on phytoplankton, which could be harmed in two ways. First, their food-making process—photosynthesis—could be diminished; second, UV-B rays could damage their genetic material and prevent them from reproducing. Scientists have found that ultraviolet rays penetrate sixty feet deep in clear water on a cloudless day but are most powerful in the top few feet—the depth to which phytoplankton rise, because they need certain wavelengths of solar energy found close to the surface for food making.

Biologists have discovered a wide range of responses by different kinds of phytoplankton to increased UV-B rays. Some are not harmed, while others are highly sensitive and do not survive. Some suffer genetic damage but then repair it and reproduce normally. Food making by phytoplankton near the ocean surface is reduced by UV-B radiation, but the loss is not as great as feared.

Since some kinds of phytoplankton are more vulnerable than others, the ozone hole over Antarctica may enable some species to thrive while others suffer. In 1991 Dr. Susan Weiler, a marine ecologist at Whitman College in Washington State, said, "Increasing ultraviolet radiation could change species abundance and variety of phytoplankton and ripple through the food chain, having effects we really can't predict."

In the Antarctic spring of 1990, biologists found that increased UV-B radiation had caused a drop of up to 12 percent in food production by ocean phytoplankton. Through 1994, however, researchers had found no evidence that the Antarctic ozone hole had caused major harm to ecosystems. Since Antarctic plants and animals are likely to be exposed to even stronger doses of ultraviolet rays in coming years, researchers continue to look for changes.

In 1994 researchers from Oregon State University announced that UV-B radiation was killing the eggs of frogs and toads in Oregon's Cascade Mountains. The eggs of these amphibians are laid in shallow water and are not protected by shells. When biologists used a filter to block UV-B rays from frog and toad eggs, the survival of the young in the eggs increased dramatically. All over the world, populations of frogs and toads have been declining, especially in mountain areas, and increased ultraviolet light is the likely cause.

Some scientists and agricultural experts are worried that thinning of the ozone shield will reduce food supplies worldwide. Biologists have learned that the eggs and young of fish and crustaceans can be harmed by strong ultraviolet rays. Botanists have exposed about three hundred plant species (or varieties of species) to increased levels of UV-B radiation. The plants tested included such major food

sources as wheat, corn, and rice. Some varieties were resistant to UV-B rays and were unharmed, but nearly two-thirds of the plants suffered cell and tissue damage and grew poorly, producing smaller harvests. Will the strong doses of ultraviolet light yet to come reduce world supplies of fishes and grains? This question will probably be answered early in the next century, when all unprotected animals and plants will be exposed to more harmful sun rays.

In the 1990s thousands of scientists all over the world continue to study the atmosphere and the threats of both vanishing ozone and global warming. (In several ways these problems are chemically interconnected. As the ozone layer grows thinner and more ultraviolet light reaches the earth, the UV light reacts with pollutants and produces more ground-level ozone, which, like carbon dioxide, is a "greenhouse" gas that warms the lower atmosphere.) Evidence supporting the ozone-depletion theory continues to mount. In 1993 Canadian scientists an-

Many species of toads, frogs, and salamanders lay eggs that are usually unprotected from increased ultraviolet light.

nounced the results of a four-and-a-half-year study that showed an increase in UV-B radiation reaching the ground. This came as no surprise to atmospheric scientists but was significant because this was the first time an increase in harmful UV-B rays had been detected by instruments outside Antarctica.

Doctors James Kerr and Thomas McElroy of Canada's Atmospheric Environment Service began measurements of ultraviolet light in Toronto in 1989. They knew that clouds and pollution can reduce the amount of this light reaching the ground, so measurements were taken at least hourly from dawn to dusk in all kinds of atmospheric conditions. They also used a highly sensitive instrument that enabled them to pinpoint and measure the most harmful UV-B radiation, which is normally blocked by ozone. They found that this radiation increased by 35 percent in the wintertime and by 7 percent in the summer. These increases coincided with losses of ozone, also measured daily, over Toronto. This important Canadian study continues, and the United States has speeded up its plan to establish a network of similar instruments to monitor changes in ultraviolet light.

Most atmospheric scientists hoped that the rapid reduction in global production of CFCs had averted an environmental disaster. Once the CFC industry grudgingly agreed with the scientific evidence, it moved quickly to find and substitute other chemicals. (See "Good-bye to CFCs," on pages 57–58.) One reason was that the costs of replacing CFCs had been greatly overestimated by economists. Besides, there were great business opportunities in making the most successful alternatives to CFCs. As a result, the CFC industry cut production earlier than required by the Montreal Protocol.

The Ozone War was over...or was it? In the mid-1990s a rearguard action is still being fought by a few scientists and some political conservatives. They continue to challenge the evidence supporting the ozone-depletion theory, and even to question the motives of atmospheric scientists. They call the theory a hoax designed to provide jobs for scientists, who are sometimes called such things as "dunderhead alarmists."

If there is any ozone depletion, these critics argue, it has natural causes: chlorine from seawater and volcanoes reaching the stratosphere. They credit Mount Erebus, an active volcano in Antarctica, as the cause of the ozone hole there, citing a scientific report that stated Erebus emits more than a thousand tons of chlorine daily.

Atmospheric scientists, geologists, and other scientists have responded with a wealth of facts. CFC molecules are inert and are not soluble (that is, they do not dissolve in water); this enables them to rise to the stratosphere and eventually release their chlorine atoms. In contrast, chlorine from natural sources is soluble; it is washed out of the lower atmosphere by rain and other precipitation, so it does not reach the stratosphere. Furthermore, Mount Erebus is not an explosive volcano, so the chlorine it emits never rises far in the atmosphere. Finally, Erebus produces only fifteen thousand tons of chlorine annually, not a thousand tons a day.

Perhaps the most disturbing thing about these continued attacks on more than two decades of scientific research is how the news media handle them. Few re-

Thinning ozone high over Antarctica cannot be blamed on the small amounts of chlorine emitted from Mount Erebus, since this chlorine never reaches the upper atmosphere.

porters and journalists are very knowledgeable about scientific matters. They are looking for a good story, one that people will want to read or watch on television. Ideally, the story will be fairly simple and clear-cut—black and white without shades of gray. A headline and a lead paragraph that exaggerate dangers are alluring to readers, but, on the other hand, a journalist should appear to be evenhanded. One way to achieve "balance" *and* make a story interesting is to present an issue as a kind of combat among experts.

This is how television news reports often treat such vital matters as the human-made threat to the ozone layer: Scientist A is pitted against Scientist B, with the viewer deciding the winner. Scientist A may be an atmospheric chemist with deep knowledge of the facts, while Scientist B may have little or no expertise, but most viewers of such "balanced" television reports do not know how unbalanced they are.

The news media have a "penchant to turn everything into a boxing match," said Dr. Stephen Schneider, an atmospheric scientist at Stanford University. Combat between opposing sides may be a good journalistic approach in reporting wars, sports, political campaigns, and court cases, but it is usually a poor way to present knowledge about issues of science and technology. So the news media bear some of the blame for delays in halting the manufacture of chlorofluorocarbons (and for continuing to fan the embers of the Ozone War). Political leaders, timid regulatory agencies, and CFC-related businesses also bear some of the responsibility.

Scientists all over the world are still trying to answer questions about the complex chemistry and dynamics of earth's atmosphere. But this book raises other questions, one of them being how people today deal with an issue like the threat to the ozone layer.

In a sense, some businesses conducted a gigantic environmental "experiment" that eventually released 750,000 tons of chlorine a year into the stratosphere. Only great efforts by some scientists, environmental groups, and others are now bringing the "experiment" to an end. The burden of proof was always on those who ques-

tioned the safety of CFCs, not on those conducting the "experiment." Surely chemicals do not have the rights of people; they aren't innocent until proven guilty. How can future "experiments" of this kind be prevented?

Whether the issue is vanishing ozone, global warming, or some other potential environmental threat, the general public and political leaders have difficulty dealing with scientific uncertainty. People want simple answers to sometimes complex questions. The process of science, however, is open to new evidence, and there is often a degree of uncertainty in its answers. How much certainty is needed in order to require actions against chemicals like chlorofluorocarbons?

As more and more harmful ultraviolet rays strike at earth's life in the next decade or two, we can be thankful that the CFC "experiment" in the atmosphere was not allowed to continue. We need to keep asking, however, why it was so difficult to stop.

Scientists are still learning about the chemistry and dynamics of the earth's atmosphere and are still searching for harmless alternatives to CFCs.

UV INDEX • FORECAST FOR JUNE 28, 1994

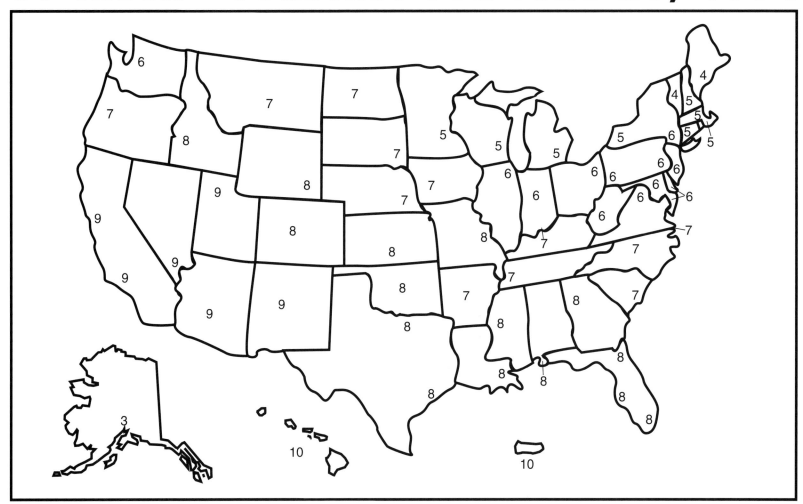

• E X P O S U R E •

MINIMAL	LOW	MODERATE	HIGH	VERY HIGH
0, 1, 2	3, 4	5, 6	7, 8, 9	10+

In 1994 the National Weather Service began issuing daily forecasts of ultraviolet exposure for 58 U.S. cities, in order to remind people to reduce their exposure to this harmful radiation. The higher the UV Index number, the greater the danger.

WHAT YOU CAN DO

People concerned about the earth's ozone layer can take action in two different ways.

One way is simply to avoid or reduce exposure to harmful ultraviolet rays, which will increase in coming years. No one is immune to harm from ultraviolet radiation, but fair-skinned people who sunburn easily are most vulnerable. Use "sun sense" to protect yourself. The single most important step is to avoid much exposure to direct sun when it is at its peak, from 10 A.M. to 2 P.M. (11:00 A.M to 3:00 P.M. daylight saving time). If possible, wear a hat and long sleeves during those hours. Apply a sunscreen lotion to exposed skin. The lotion should have a sun protection factor (SPF) of at least 15. Apply the lotion twenty minutes before exposure to full sunlight and reapply every two hours and after swimming or vigorous play or work.

Beware of the notion that sunscreen lotions make long exposure to sunlight safe. Most of them block UV-B rays that may cause certain kinds of skin cancers. Despite advertising claims, however, most sunscreens do not block UV-A rays. While these rays do not cause sunburn, there is some evidence that UV-A radiation may cause melanoma, the most deadly skin cancer. Only a few expensive sunscreens with the ingredient Parsol 1789 offer protection from both UV-A and UV-B radiation.

In addition to protecting yourself from ultraviolet rays, you can learn more and stay informed about the ozone-depletion issue by checking recent issues of the periodicals recommended on page 62. Environmentalists are still trying to stop production of methyl bromide, a pesticide that releases ozone-destroying bromine. You can learn about this issue, and about actions to take, by writing to the government agency and environmental groups listed below.

Environmental Protection Agency
401 M Street, SW
Washington, DC 20460

Natural Resources Defense Council
1350 New York Avenue, NW
Washington, DC 20005

Environmental Defense Fund
257 Park Avenue South
New York, NY 10010

GOOD-BYE TO CFCS

In 1988 about 2.1 billion pounds of chlorofluorocarbons were produced globally. By the mid-1990s production of CFCs had been cut by more than half, and many uses were being phased out. In some cases manufacturing processes were changed so that CFCs were no longer needed; in others a substitute for CFCs was found.

CFC-113 has been used for cleaning computer circuit boards and other electronic equipment. It represented 17 percent of all CFCs produced each year. The use of CFC-113 was reduced in part by soldering circuit boards with materials that did not require cleaning by CFCs. Substitute cleaners, including a compound made from citrus-fruit rinds, were also found.

Other CFCs (CFC-11 and CFC-12) were used in making rigid and flexible plastic foam for insulation, drink containers, and packaging. In the United States, substitutes for all of these uses were adopted by 1995. However, finding substitutes for the CFC-11 and CFC-12 used in refrigerators and air conditioners was more difficult. The first substitutes developed were hydrofluorocarbons (HFCs) and hydrochlorofluorocarbons (HCFCs). The former contain no chlorine that could harm ozone. HCFCs contain chlorine, but since they also contain hydrogen, these compounds break down in the lower atmosphere so that little of the chlorine can reach the ozone layer. However, these substitutes are far from ideal. Because they use more energy, they are less efficient than CFCs in refrigerators and air conditioners. They are also heat-trapping gases that will add to the problem of global warming. With engineers and chemists all over the world searching for better refrigerants, it seems likely that HFCs and HCFCs will be only a temporary solution until new, efficient, and nonpolluting substitutes are developed.

The Du Pont company had planned to end production of CFC-12 in 1994 but was asked to continue for a time by the Environmental Protection Agency. The reason: Ninety million older cars had air conditioners that could not easily be adapted to use the HFC refrigerant that was being used in the modified air conditioners of brand-new cars.

At the beginning of 1994, industrialized nations stopped making halons, gases used in fire extinguishers. Developing nations were given another decade to phase out their production. Like CFCs, halon molecules are long-lasting and rise to destroy ozone in the stratosphere. Chemists doubt that they will find a substitute as effective as halons; instead, they hope to develop several substitutes that meet different fire-extinguishing needs.

GLOSSARY

atom—the smallest unit of an element (such as oxygen or carbon), consisting of a central nucleus surrounded by orbiting electrons.

cataract—a clouding of an eye's lens that results in blurred vision. Cataracts can be caused by exposure to strong ultraviolet (UV-B) light.

chlorofluorocarbons (CFCs)—compounds made of chlorine, fluorine, and carbon that were widely used as propellants in spray cans, as coolants in refrigerators and air conditioners, and in other ways until scientists found evidence that chlorine from these compounds was destroying the stratosphere's ozone layer. Millions of tons of long-lasting CFCs remain in the atmosphere and will continue to harm the ozone layer for decades.

ecosystem—all of the living and nonliving parts of a given place in nature; for example, a coral reef or a desert.

global warming—an increase in the atmosphere's temperature caused by human activities that add carbon dioxide and other heat-absorbing gases to the atmosphere.

"greenhouse" gases—carbon dioxide, methane, and other gases that absorb heat energy from the sun and help warm the earth's atmosphere. Life could not exist on earth without this greenhouse effect, but increasing amounts of these gases can raise the global temperature and harm plant and animal life.

hydrocarbons—chemical compounds made of hydrogen and carbon atoms that are given off during such natural processes as decay and also when gasoline, coal, and other fuels are burned.

melanoma—skin cancer; the most serious, malignant melanoma, is often fatal. It is caused by exposure to strong wavelengths of ultraviolet light from the sun.

molecule—the smallest possible amount of a compound that still has the physical and chemical characteristics of that substance. A molecule of ozone consists of three oxygen atoms.

ozone—a form of oxygen present in the earth's atmosphere in small amounts. In the lower atmosphere, ozone is a pollutant that damages materials and harms plants and animals. In the upper atmosphere, a layer of ozone makes life on earth possible by shielding the surface from most of the sun's harmful ultraviolet rays.

photosynthesis—the manufacture of food, mainly sugar, by green plants. The process combines carbon dioxide and water in the presence of chlorophyll, using solar energy and releasing oxygen.

plankton—tiny drifting plants and animals that live in salt and fresh water. The plants, or phytoplankton, are the first link in many aquatic food chains. The animals, or zooplankton, feed on phytoplankton.

smog—this word was coined in London, England, where pollutants from *smoke* mixed with *fog*. Now the term is applied to a mixture of pollutants, including ozone, that accumulates in a basin or valley when atmospheric conditions keep it from dispersing.

spectrophotometer—a device that measures the full range, or spectrum, of sunlight, including ultraviolet radiation. Since ozone blocks most ultraviolet light, observations by spectrophotometers are used to detect changes in the amounts of ozone present in the stratosphere.

stratosphere—the upper atmosphere, between seven and thirty miles above the earth's surface, which holds the earth's life-protecting ozone layer.

troposphere—the lower atmosphere, which extends to about seven miles above the earth's surface. Nearly all of the atmosphere's weather occurs in the troposphere.

ultraviolet radiation—invisible rays of energy from the sun that have shorter wavelengths than visible violet light. The shorter the wavelength, the more harm ultraviolet rays can cause, including sunburn, aging of skin, and skin cancers.

FURTHER READING

Since our understanding of the ozone layer has grown with time, the more recently published books and articles listed below are the most useful. For even more up-to-date information, see recent issues of *Science* and *Discover,* which are available in many libraries. The British publications *Nature* and *New Scientist* are also good sources that can be found in some libraries. In all periodicals, be sure to read the letters to the editor section. Watch for letters that comment on previous articles; they often shed light on disagreements among scientists, and on the economic and political factors that are intertwined with scientific controversy.

Blaustein, Andrew. "Amphibians in a Bad Light." *Natural History,* October 1994, pp. 32–39.

Cagin, Seth, and Philip Dray. *Between Earth and Sky: How CFCs Changed Our World and Endangered the Ozone Layer.* New York: Pantheon, 1993.

Fisher, Marshall. *The Ozone Layer.* New York: Chelsea House, 1992.

Fishman, Jack, and Robert Kalish. *Global Alert: The Ozone Pollution Crisis.* New York: Plenum Press, 1990.

Kerr, Richard. "Ozone Takes a Nose Dive After the Eruption of Mt. Pinatubo." *Science,* April 23, 1993, pp. 490–91.

Nance, John. *What Goes Up: The Global Assault on Our Atmosphere.* New York: William Morrow & Company, 1991.

Roan, Sharon. *Ozone Crisis: The 15-Year Evolution of a Sudden Global Emergency.* New York: John Wiley & Sons, 1989.

Stolarski, Richard. "The Antarctic Ozone Hole." *Scientific American,* January 1988, pp. 30–36.

Taubes, Gary. "The Ozone Backlash." *Science,* June 11, 1993, pp. 1580–83. (See also a response to this article and a reply by Dr. Sherwood Rowland in the letters department, pages 1101–3, of the August 27, 1993, issue.)

INDEX